The United States National Climate Assessment

NCA Report Series, Volume 5b

Monitoring Climate Change and its Impacts:
Physical Climate Indicators

March 29-30, 2011
Washington, DC

National
Climate
Assessment

U.S. Global Change Research Program

NCA Report Series, Volume 5b

Monitoring Climate Change and its Impacts: Physical Climate Indicators

NCA Report Series

The National Climate Assessment (NCA) Report Series summarizes regional, sectoral, and process-related workshops and discussions being held as part of the third NCA process.

This workshop focused on monitoring changes in the physical climate system, as part of a series that includes workshops on the impacts of climate change on ecosystems and socio-economic systems. The workshop was held in Washington, D.C. on March 29-30, 2011. Volume 5b of the NCA Report Series summarizes the discussions and outcomes of this workshop. A list of planned and completed reports in the NCA Report Series can be found online at http://assessment.globalchange.gov.

CONTENTS

Executive Summary

Workshop Details

The National Climate Assessment (NCA) convened a workshop on "Physical Climate Indicators" from 29 to 30 March, 2011 as part of a series on "Monitoring Climate Change and its Impacts". The overarching goal of this workshop was to identify a few broad categories of potential physical climate indicators using a set of priorities developed by the NCA, and to provide a clear justification for how they would inform the Nation about climate change. Additional goals included providing input on the overall NCA framework for selecting the indicators and suggesting methodologies to construct indicators. Although one of the workshop goals was to address the status of current observational networks to support indicators, this was not a main focus of any single discussion.

The nearly 60 participants, primarily from federal agencies, received a white paper in advance of the workshop that detailed the NCA vision for a coordinated suite of climate-related physical, ecological, and societal indicators. The intent of these "national indicators of change" is to develop a way to evaluate and communicate over time both the rate of change in impacts and the capacity to respond to climate drivers. These indicators will be tracked as a part of ongoing, long-term assessment activities, with adjustments as necessary to adapt to changing conditions and understanding. An initial framework was provided to workshop participants to ensure that everyone understood the proposed audience, scope, and purpose of the indicators. A common lexicon was defined since indicator terminology varies widely. In addition, several potential approaches to grouping the indicators were presented.

Workshop Structure

Participants spent most of their time in small breakout groups with facilitators, working to address a common set of questions. The workshop was structured to start with the broadest issue and then focus down as the workshop progressed. The first breakout therefore solicited comments on the NCA indicator framework, followed by discussion of the potential approaches to organizing the physical climate indicators. Once several approaches were identified, the groups then worked to define specific measurements, or types of measurements, that could be used to create the indicators. The overall goals,

purpose, audience, and scope received a wide range of comments from the four breakout groups.

Comments on NCA Physical Indicator Vision and Framework

- There was concern about the ability of the indicators to function at both the national and the regional to local level. Several comments were made regarding the ability of the NCA to be, as one participant said, "everything to everyone." Obviously, the NCA must strike a balance between providing too-detailed information at regional/sectoral levels at the expense of useful background information on a national scale.

- Questions remained about whether indicators should address climate change impacts exclusively, or whether they should also include vulnerabilities and/or drivers of climate change. Clarifying these aspects will be critical to the success of the indicator suite both as a communication tool and to be useful for decision makers.

- Participants were clear that leveraging the availability of existing resources would be necessary to meet the required goals of the NCA as an ongoing, sustained effort. Several of the breakout groups noted a need to inventory the various indicator-based assessment efforts. As pointed out by one participant, the development of these indicators will likely rely heavily upon federal agencies; therefore it should be determined what is already at their disposal.

- Indicators must be developed using high-quality data sources. Indicators must satisfy the constraints of credibility and transparency. Workshop participants agreed that data sources should be well documented and peer reviewed to the extent feasible. Using only the highest quality data sources may restrict which metrics are available for the construction of the indicators.

- Communicating the indicators will require careful consideration. A number of presentations and breakout discussions expounded on the importance of crafting how to communicate indicators. The NCA should engage communication specialists and graphic artists early in the process. Feedback needs to be actively sought from non-scientists on the intelligibility of the indicators to the general public.

- Efforts must be made to relate climate change and impacts in a manner that people can readily perceive. To communicate effectively with a broad audience, the NCA will need to go beyond the use of basic variables. Though indicator categories may be based on basic metrics of temperature and precipitation, it could be more useful to quantify derivatives from these metrics such as growing season length or number of days greater than some temperature threshold.

- Finally, the workshop participants noted the need to integrate the physical indicators with the ecological and societal indicators. Once reports from all three workshops are complete, an obvious next step will be to blend the report findings and knowledge gained into a revised vision for an integrated suite of ecological, physical, and societal climate indicators that meet the needs of the National Climate Assessment. .

Grouping of Indicators

The workshop white paper provided three example approaches to group the physical climate indicators: 1) statistical, 2) thematic, and 3) physical system, with potential indicators for each approach. The breakout groups discussed these approaches and were challenged to consider alternatives. In general, the participants gravitated towards a mixture of the approaches, with an emphasis on thematic indicators that could readily include statistical measures. Two analogies for the indicators were presented and discussed in that context – warning lights and vital signs. There was concern that framing the indicator suite as warning lights might prejudge problems and imply set thresholds, whereas vital signs would be seen as more consonant with indicating the current state of the climate system and potential impacts.

Indicator Categories

Though following different approaches, several common indicator categories were proposed from the breakout groups. These included

- Extreme Events/ Natural Hazards;

- Biophysical Changes;

- Hydrological/Freshwater;

- Timing;

- Coastal; and

- Cryospheric Changes.

The breakout groups were split on the need to include an indicator category encompassing anthropogenic forcing. Most of the proposed indicator categories are impacts-oriented. The workshop participants generally found indicators that were supportive of adaptation and mitigation to be more compelling than indicators that simply track the past, present, and projected changes of a climate variable. For each of the proposed categories, a number of important impact-related subtopics were identified. It remains an outstanding issue to determine how best to incorporate this information into a parsimonious indicator that still encompasses important climate change impacts, vulnerability, and adaptation.

National Climate Assessment

The United States Global Change Research Program (USGCRP) has been tasked with providing a coordinated strategy and implementation plan for assessing the changing climate and potential impacts on the Nation. This strategy is being developed with the intent to both provide critical support to the third National Climate Assessment (NCA) and establish a mechanism for an ongoing assessment capability. The objectives of the next NCA report, due in the summer of 2013, are stated in Box 1. It is expected that an ongoing NCA process will be established and sustained through a cooperative community-wide effort that incorporates a multitude of federal, state, and local governmental agencies, non-governmental organizations, and private interests. The NCA process should lead to enhanced coordination of the various climate assessment efforts and create a strong linkage between stakeholders and data providers. Once an ongoing assessment capability has been established, it is expected that assessment reports will be produced on a regular basis, with less emphasis on producing a single major report every 4 years as required in the Global

Change Research Act. Additional emphasis has been placed on education and communication in the NCA framework. This includes the development of a web-based portal for the deployment of NCA information and supporting material.

The National Climate Assessment Development and Advisory Committee (NCADAC), a federal advisory committee (FAC), is charged with conducting the Assessment. It is comprised of approximately 60 individuals from academia, the non-profit sector, industry, and the federal government. The NCADAC will examine a variety of inputs from scientists and regional and sectoral assessment data and reports to build assessment products. They are also assigned to provide guidance to the ongoing NCA process.

The current NCA approach[1] differs in multiple ways from previous U.S. climate assessment efforts, being more focused on: 1) supporting the Nation's activities in adaptation and mitigation and on evaluating the current state of scientific knowledge relative to climate impacts and trends, 2) a long-term, consistent process for evaluation of climate risks and opportunities and providing information to support decision making processes within regions and sectors, and 3) establishing a permanent assessment capacity both inside and outside of the federal government. The NCA will therefore be an ongoing process that draws upon the work of stakeholders and scientists across the country. Assessment

---

[1]The current NCA strategic plan can be found at: http://www.globalchange.gov/what-we-do/assessment/backgroundprocess/strategic-plan

activities will support the capacity to do ongoing evaluations of vulnerability to climate stressors, observe and project impacts of climate change within regions and sectors, allow for the production of a set of reports and web-based products that are relevant for decision-making at multiple levels of space and time, and develop consistent indicators of progress in adaptation and mitigation activities. Products of the NCA process should be useful within management and policy contexts.

A Set of Indicators for the Climate System

The National Climate Assessment anticipates the development of a set of indicators that can be used to assess the changing climate and its impacts at a national scale. The physical climate indicators discussed in this workshop are being developed as part of a coordinated suite of climate indicators that also includes ecological and societal indicators. The NCA indicators are meant to: 1) provide meaningful, authoritative climate-relevant measures about the status, rates, and trends of key physical, ecological, and social variables and values to inform decisions on management, research, and education at regional to national scales; 2) identify climate-related conditions and impacts to help develop effective adaptation measures and reduce costs of management; and 3) document and communicate the dynamic nature and condition of Earth's systems and societies, and provide a coordinated benchmark for all regions and sectors. These indicators will help track climate change, whether natural or anthropogenic, and should serve as an effective communications tool. A goal of these indicators is to move from the traditional existing indicator efforts, drawing upon their successes and failures, and support improved communication of how the climate is changing in a manner that high-level decision makers find useful and comprehensible. It is important to provide an integrated (e.g., physical, ecological, societal) view of the changing climate and its impacts to better inform management and policy actions that can be targeted towards successful mitigation and adaptation activities.

Some proposed qualities of physical indicators set forth within the NCA framework are:

- Based on physical observations;

- Related to climate impacts;

- Easily communicated to a broad audience;

- Relevant to multiple sectors;

- Based on available, high-quality data;

- Applicable to decisions at multiple scales;

- Able to be aggregated from local to national scales; and

- Indicator components can be projected into the future using models and tools that currently exist.

Participants in the workshop were asked to keep these requirements in mind while recommending the approach to indicators. The participants also provided input on the value and clarity of the requirements.

Workshop Motivation and Structure

The primary aim of the physical climate indicators workshop was to identify a set of broad categories from which the physical indicators could be constructed for the NCA. Ultimately, it is expected that approximately five to eight indicator categories will be selected to provide information on physical aspects of the changing climate system and impacts. In addition to identifying these broad categories, participants were asked to provide input on how to construct the indicators and provide a clear justification of how those indicators would inform the Nation. Comments were also sought on the clarity of the concept and purpose of the overall NCA framework including the criteria for selecting indicators and how they can best be implemented (data availability, labor effort, etc). The charge given to workshop participants from the workshop white paper is given in Box 2.

Box 2: Workshop Charge Given to Participants

Participants are asked to:
1. Comment on elements of the general NCA framework;
2. Identify broad physical climate categories that could be used in the NCA report (i.e., that fit the framework);
3. Provide suggestions of candidate metrics that inform each objective; and
4. Assess the feasibility of using the identified indicators given the current data sources available and note what additional resources would be needed to implement these indicators.

Workshop participants were invited from federal agencies, non-governmental organizations, academia, and industry in an effort to provide multiple viewpoints on the NCA framework and indicator selection process. This workshop was organized in a similar manner to previous NCA workshops with a mix of plenary talks with question and answer sessions, and breakout group discussions. The plenary talks focused on potential applications of climate indicators and previous indicator efforts. At the previous NCA ecological indicators workshop, held November 30-December 1, 2010, a bottom-up approach was used wherein participants started with available data, observations, and metrics in an attempt to integrate these into broad-scale indicators. For the physical climate indicators workshop, an alternative approach was suggested to address the high-level "strategic" issues first, leaving data and metrics for later discussions. There were four breakout sessions that each addressed the same breakout charge. The breakout charge included commenting on the overall NCA approach, developing broad-scale indicator categories, and finally discussing what data/metrics would support those indicator categories. The full workshop agenda can be found in Appendix A.

Previous Indicator Efforts

A number of efforts within the climate science community have developed indicators to track and to communicate relevant aspects of the changing climate system. These indicator efforts often use similar basic data metrics, but they have been packaged in differing ways to respond to the objectives of particular agencies and entities. Indicators from these efforts may or may not be precisely aligned with those of the National Climate Assessment. As background for the workshop participants, a number of these endeavors were reviewed in a pre-workshop white paper[2] and lessons learned were presented during the workshop by those familiar with specific indicator efforts. These indicators were discussed within the context of the NCA indicator requirements and vision. Taken together, these and other efforts provide a repository of experience that can be used to help design the NCA indicator framework and produce an integrated set of indicators capable of effectively communicating information to a wide audience.

---

[2]A copy of the workshop white paper and presentations can be found at: https://sites.google.com/a/usgcrp.gov/physical-indicators/home

Prior to the breakout sessions, two plenary panel
discussions provided an overview of how indicators
have been developed and used in the past. A list
of the presentations can be found in Appendix A,
and four main themes that appeared throughout the
presentations are provided in Box 3. A summary of
the main points raised by the presenters and plenary
discussion are provided below.

The first panel discussion focused on the potential
application of climate indicators to the issues of
concern to a variety of stakeholders. The presenters
attested to the advantage of indicators for communicating large amounts of climate information
in a useful way. However, cautionary tales of the
inappropriate uses of indicators and indicators that
did not meet the intended needs were presented.
For example, some insurance companies have attempted to factor in correlations between long-term
changes in land-falling hurricanes and the Atlantic
Multidecadal Oscillation when setting premiums.
This approach encountered stiff resistance from state
legislators who felt contrasting results in the scientific literature provided too little support for adjusting
insurance premiums, placing undue burden (raised
premiums) on a significant number of people. Both
concern about and support for developing indicators relative to thresholds or tipping points was
provided. It was noted by presenters that effective
application of indicators relies on the ability to
translate science into the currency of stakeholders. For instance, changes in temperature can be
expressed as changes in growing season length for
agricultural stakeholders or changes in snowmelt
patterns for water managers.

The second panel discussion consisted of four
presentations aimed at familiarizing the workshop

attendees with previous indicator efforts and lessons learned. One common lesson identified was
the difficulty in predicting how climate indicators will be used; the audience is dynamic. As an
example, global mean temperature indices were
developed for the science community, but now
they are reported to the press. The Palmer Drought
Index was developed for agricultural stakeholders,
but other sectors and the general public now use
it as well. Several presenters also noted the need
for indicators to have traceability to high-quality
observations. Designing effective physical indicators
of climate has the potential to help people visualize
climate change in a way that they can incorporate
it into their work and plans. Thus, in addition to
being scientifically sound, the physical indicators
of climate need to be understood by (or readily ex-
plained to) people without formal training in climate
science. Based on previous experiences, maps are
more readily interpretable by the public than x-y
plots conventional in the science community. The
presenters generally found the development of a
successful indicator required properly targeting its
need or purpose, the audience, and scope.

Stakeholder involvement throughout the process,
and proactively soliciting user feedback on the
relevance of indicators, will help ensure that indicators are used and useful. Even if an effective suite of
physical indicators of climate is developed, broad
communication mechanisms will be required to
guarantee the information is disseminated. Presentations on the IGBP (International Geosphere-
Biosphere Programme) Climate Index and NOAA's
Annual Greenhouse Gas Index provided direct evi-
dence of the need to carefully consider how to present indicators. A message that resonated throughout
the discussions was that while scientists may be
experts in their field, they tend to be bad judges
of what non-scientists find easy to digest. Engag-
ing communications experts early on, and testing
communications strategies with a wide audience,
can help produce indicators that will be useful for a
broad community.

NCA Indicator Vision

The NCA vision for indicators is a small (<20),
coordinated suite of climate-related physical, ecological, and societal indicators that both takes the
pulse of key aspects of climate and climate impacts
for the United States and is easily communicated to
interested parties. These indicators will be tracked
as a part of ongoing, long-term assessment ac-

tivities, with adjustments as necessary to adapt to changing conditions and understanding. During the first breakout session, workshop participants were separated into four breakout groups and charged to discuss the NCA vision (*i.e.*, purpose, scope, audience) using the following questions as a guide:

- Purpose: Is the proposed purpose for the NCA indicators clear, complete and compelling, and what changes might improve these?

- Scope: How should the physical climate indicators relate to regions or sectors for the NCA? How should the physical climate indicators relate to ecological and societal indicators of climate change also being developed for the NCA?

- Audience: Are the potential user groups of the NCA indicators appropriate, or too narrow or broad? Do the user group's needs and indicator purpose align?

Feedback and discussion in the breakout groups clustered around three themes: (i) purpose and scope of indicators; (ii) indicator development, integration of data, and regional/sectoral considerations; and (iii) coordination with other agency efforts.

Purpose and Scope of Indicators

The workshop participants discussed potential applications of physical climate indicators and felt that indicators offer an opportunity to influence public discourse on climate change, particularly if the indicators have wide relevance to and resonate with the public. It was noted by several breakout groups that once the indicators are created, they might be used for a number of unintended purposes. Likewise, it is also possible that the broader community may not use indicators developed by the NCA even if they are understandable and informative. From a federal perspective, indicators can help interagency efforts in strategic planning discussions and the identification of vulnerabilities. Indicator selections may help identify and set budget priorities. It was pointed out that the best approach was to provide indicators that are first and foremost transparent and comprehensible, and secondly are sustainable over the long run – and therefore more likely to be used appropriately in a planning process.

In addition to addressing climate impacts, workshop participants discussed whether and how indica-

tors could address adaptation. They had conflicting viewpoints on the issue. Some felt the goals for the indicator framework should clearly include adaptation and that indicators need to have a way to feed into adaptation and mitigation decision making if they are to be useful. However, some participants felt that NCA physical climate indicators would not necessarily be used in cases where adaptation and mitigation activities are already taking place, and that the NCA indicator vision should soften its requirement of including adaptation.

The NCA indicator vision presented during the workshop called for specific uses of indicators. However, some workshop participants commented that the vision should not presume the usages of indicators. Rather, some felt the goal should be to convey information about the climate, and then decision makers would decide relevance. The breadth of viewpoints, and at times confusion among participants, regarding how indicators will or should be used may reflect a weakly-stated vision for the indicators. The need for the NCA indicators to be used in decision making should be clarified. The vision for indicators needs to match the vision for the overarching NCA approach and it is not yet clear whether the Assessment is only about assessing the state of the climate, or whether it encompasses climate impacts, vulnerability, and adaptation as well.

Indicator Development, Integration, and Regional/Sectoral Considerations

A significant topic of discussion was the difficulty or advisability of creating "one size fits all" indicators that are useful given the different regional interests and sectoral backgrounds of the stakeholders. Some workshop participants felt that "providing a coordinated benchmark for all regions and sectors" was too ambitious. Not all indicators in a suite will give a consistent or meaningful signal, nor should indicators be chosen on that basis. An example is that ice storms are defined differently for different applications. There are different thresholds for "ice storm" in different sectors (how much ice before it is defined as an ice storm for transportation, energy, *etc*). A counter point offered was that the NCA could create an indicator, such as timing of seasonal events that can be calculated on a national basis, but linked to regional observations (*e.g.*, timing of first freeze in one region *vs*. timing of maximum runoff in another). This type of indicator could be adapted in different sectors and regions and related to the needs of local stakeholders. In this way, indicators could be synthetic yet traceable back to regional concerns. Many comments suggested that there is value both in being able to customize or drill down to the regional scale and having core national-scale indicators.

Related to this was discussion about whether indicators should focus on the "state of the climate" (status) or policy relevance. One view was that the indicator system is separate from the Assessment, and that indicators should be more for policy than science. A challenge for the climate science community will be to help policy, planning, and decision makers determine what information is pertinent to their specific questions. Having a sense of what kinds of questions that are asked by the policy audience could then help clarify useful indicators. This could be accomplished by exploring the questions that regional stakeholders and policy makers are being asked. Some workshop participants felt policy relevance should be emphasized to a greater degree in the indicator vision. With an emphasis on policy relevance, however, five to eight physical indicators might be too few to address sectoral and regional questions. It was noted that it will not be feasible for the NCA to have indicators that address every individual sector and region, and that instead, the NCA should take ownership of a small number of indicators that could serve as an umbrella that enables others to drill down to sectors and regions with more tailored analyses.

The initial requirement that the indicators "Readily communicate integrated information" led to a discussion of whether that would require the integration of many variables into a composite index; if so, is the creation of composite indices desirable? Some concerns were expressed with composite indices: (i) there can be interdependencies that reinforce or cancel out indicator changes, and (ii) it can be difficult to link the indicator to vulnerabilities. A state variable (*e.g.*, temperature or precipitation) or single direct measurement linked to physical principles may be more useful to stakeholders in terms of clarity. Support for data integration to form a composite index was also expressed, however. The advantage of a composite index is that it would provide a synthesis of the abundance of numerous scientific measurements from disparate sources and therefore could be a good communication tool. That is, several indicators could be integrated to tell a compelling story that conveys what physical changes are occurring and to what the public should react.

Other specific comments regarding the indicator category development include

- In the development of indicators, transparency and traceability of the data are important. A common-source data set can be important.

- The way in which the NCA will discuss and deal with climate variability *vs.* climate change needs to be clarified as part of the indicator development.

- The indicator goals should include integration of the physical, social, and ecological indicators.

Coordination with Existing Agency Indicator Efforts

Discussions during the breakouts also examined the relationship between the NCA indicator effort and existing indicator efforts. Given the number of existing indicator/assessment efforts, some felt it might be best to begin by assembling a list of available indicators from which to work rather than starting afresh. One group noted there are significant ongoing assessments and projects that include indicators that might be useful to the NCA. The NCA could provide periodic, consistent synthesis of existing indicators without recreating ongoing work. To this end, the NCA could:

- Leave intact individual indicators for different applications as some users only require certain information and combining everything into one number is not always the most useful;

- Keep intact individual source data; and

- Conduct periodic meta-analysis to have better cross-comparison of trends.

An example of this is Environmental Protection Agency's (EPA) 2010 assessment report, Climate Change Indicators in the United States[3] that uses NOAA's greenhouse gas index and takes assessment information from the U.S. Department of Agriculture (USDA) Forest Service to produce a national synthesis that follows Intergovernmental Panel on Climate Change (IPCC) guidelines. It was noted, however, that the NCA hoped to derive a more limited list

---

[3] Available at: http://www.epa.gov/climatechange/indicators.html

than the EPA effort. The NCA could, however, develop similar guidelines through this process and develop a synthesis. Individual agencies are already monitoring climate; it is the cross-comparison of the various indicators that is lacking.

Developing a Set of Physical Indicators for the NCA

In preparation for discussing the categories of indicators that could be used in the NCA, a lexicon (Table 1) was provided in the workshop white paper in an attempt to provide a common base for discussion amongst the workshop participants who represented a variety of backgrounds. The table defines the category of the indicator, the indicator, and data metrics. The components were defined through analogies to common indicators: i) the Dow Jones or S&P 500, ii) an extreme events indicator, and iii) a hypothetical composite timing indicator. The workshop participants were asked to identify indicator categories by trying to formulate top-level questions of the purpose and goal of each indicator and then determine what indicators and data are available (or could be created) to answer those questions.

Purpose, Goal or Category for an Indicator	How is the stock market doing?	Are 'extreme' events changing due to climate?	Has the timing of physical factors shifted
Indicator	Dow Jones or S&P 500	NOAA's Climate Extremes Index	Climate Timing Index (hypothetical)
Metric/variable that supports indicator	Basket of stock prices	Annual mean of extremes related to temperature, precipatation, and drought	Change in date of: 1st frost, lake monomixis, peak stream runoff, etc.

Table 1. Table presented to workshop participants to establish a baseline in terminology. Major distinctions were made between the category of an indicator, the indicator itself, and data metrics that would support the production of an indicator. The first column provides an economic example and the second and third columns provide examples of physical climate indicators.

There are likely a limited number of metrics/variables that will meet the NCA's requirements, so a central issue for this workshop was to address how the indicators could be grouped in ways that are informative and meaningful for the targeted audience. Rather than having a disparate set of indicators, the question was posed as to whether there were any common approaches that could be used to sort the types of indicators. The white paper set forth three potential approaches that could be used for grouping the indicator categories (Box 4). The approaches suggested were: 1) an integrated statistical approach; 2) a thematic approach; and 3) a physical or systems-based approach.

The *statistical* approach includes the use of common measures to evaluate changes in a number of data metrics. These categories would include mean states, timing of events, changes in variance, and the frequency of occurrence of extreme events among others. Within a category such as mean states, there might be several components such as sea level, temperature, soil moisture, *etc* The *thematic* approach organized categories of indicators around a particular topic such as the water cycle, hazards, coastal issues, sea-level effects, and temperature effects (e.g., ski days, growing season length). A separate approach, though related to the thematic grouping, would be to develop categories relating to the *major divisions of the physical climate system*. This would entail producing an indicator describing changes in the atmosphere, coasts and oceans, the cryosphere, hydrology (e.g., humidity, precipitation, lake area), and possible climate drivers such as greenhouse gases or albedo. The distinction between the thematic and physical approach is perhaps not as strong as that between the statistical and thematic approach. The intent of suggesting these approaches was to provide the participants with a few examples of how different indicator categories may be grouped together to provide a robust sense of climate change and its impacts. The participants were reminded that these were only suggestions meant to help encourage discussion; they were welcome to propose alternative approaches or to combine the approaches.

Vital Signs vs. Warning Lights

Additional to the set of approaches proposed in the white paper, two conceptual viewpoints of indicators were presented to the workshop participants. These two frameworks, 'vital signs' and 'warning lights', were set forth to give additional focus on

how the physical indicators of climate could be regarded (Figure 1). A 'vital signs' approach has a medical analogy in which a physician will monitor a variety of types of information (e.g., weight, height, pulse rate, blood pressure, body temperature, respiratory rate) as key indicators of a patient's health and then interprets this information to assess vulnerability and critical thresholds given a patient's age, sex, and ethnicity. Taking a 'vital signs' approach, physical indicators of climate would be used to monitor key generic attributes of the climate system and these indicators would then be interpreted to assess vulnerabilities and critical thresholds given the region, sector, or planning horizon. The indicators being monitored would not necessarily provide warning, but through value-added interpretation may, either individually or collectively, signal the approach of a critical threshold or an unacceptable consequence.

Warning Lights

Vital Signs

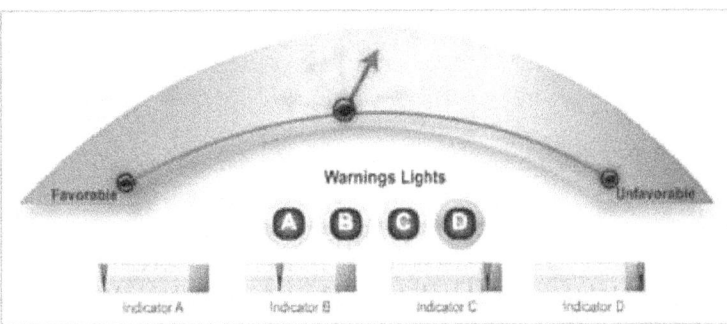

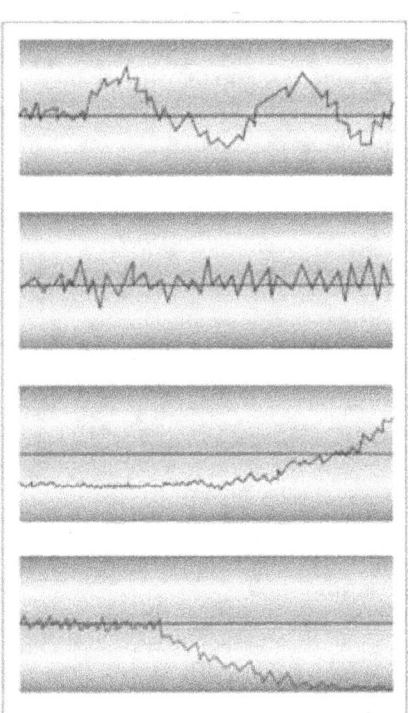

Figure 1. Two visions of indicators suggested at the meeting were 'warning lights' and 'vital signs.' The warning lights (left panel) appear on a dashboard along with a speedometer, and the lights turn on once their value exceeds a pre-determined threshold. The inverted triangle along the slider bar for each 'indicator' denotes the current status, while the colored vertical bar designates the threshold to turn the light 'on'. In this illustration, indicator D has exceeded the threshold and hence the light is red. The position of the speedometer along its track from 'favorable' to 'unfavorable' indicates the current status of the index comprising the four indicator metrics, and the position of the arrow, potentially from -180° to +180° captures the recent trend of the index over some period of time. Currently, the index is situated slightly past the 'favorable' status and the arrow indicates a trend continuing towards the right over the recent past. The vital signs illustration (right panel) depicts the time history of the same indicator (A-D) as in the left panel. The colors provide an indication of whether the indicator has crossed from a favorable (green) status to a more negative one (yellow) and finally into an unfavorable zone (red).

A 'warning light' approach has an automotive dashboard analogy in which the automotive maker has implemented a system to monitor a number of functions (*e.g.*, oil pressure, coolant temperature, volts, tire pressure) as key indicators of performance and then provides multiple warning lights that operate independently of each other to identify specific problems. Taking a 'warning light' approach, physical indicators of climate would monitor specific attributes of the climate system and when critical regional or sectoral vulnerabilities or thresholds were being approached or being crossed, a warning would be provided. In both the dashboard and climate warning light approach, the critical threshold or an unacceptable consequence is identified in advance and a warning is signaled when critical values are reached. Although posed as alternative frameworks to explain the context and role for physical indicators of climate, more in-depth discussions during the breakout session suggested that 'vital signs' and 'warning lights' may both be complementary and useful mechanisms for communication.

Properties of Good Indicators

The NCA has proposed a set of desired characteristics for indicators that were discussed among the participants throughout the breakout sessions, and these discussions are synthesized here. There seemed to be consensus about the usefulness of several of the indicator properties though issues did arise with the semantics of several criteria. The ability to skillfully project indicators from climate models was not suggested to be an eligibility requirement, though it was considered a beneficial characteristic. Several groups proposed the condition that indicators should be societally relevant. That is, they must relate to the proposed audience. It could be argued that "relating to impacts" may itself encompass this characteristic. Another suggested property was that physical indicators must be relatable to the societal and ecological indicators in some manner. For example, a physical indicator on the number of lightning strikes could be related to an ecological indicator about wildfires and to a societal indicator on damages associated with wildfires. The availability and quality of data sources was stressed as an important requirement for any indicator. It was suggested that having long records, a property allowing the ability to track historical trends, may be a prerequisite for indicators. It should be noted that including needs such as long, high-quality records and the ability to be spatially representative

may place strong constraints on what basic data metrics are available for constructing indicators.

The Aggregation of Indicators

The topic of aggregating the indicators caused confusion related to the indicator vision for the NCA. As pointed out by one group, there is an important conceptual distinction between an EPA-like approach in which you have several indicator categories under which there may be a number of stand-alone indicators, and an IGBP-like approach in which various climate indicators are merged together into a single index. Several groups noted concerns with aggregated indices, particularly with respect to communicating to the public. Participants noted that the key to any aggregation of indicators must be that its meaning will be unambiguous to the end user. Several participants felt some indicators lend themselves to aggregation better than others. Participants cautioned that if the NCA chooses to aggregate indicators into indices, it must be done in a manner that does not dilute the message of the indicator.

As a complement to the issue of aggregation, workshop participants questioned how to construct an indicator. For example, an indicator may be calculated as one number for the Nation or it could be more useful for the indicator to be scaled regionally. If nested, this would imply the development of indicators that make use of data from a local level and are relatively well-sampled spatially. Constructing indicators in this manner carries other problems. For example, the definition for "heavy" precipitation will vary geographically. However, it was noted that it is important for indicators to be traceable to at least the regional level.

Breakout Group Outcomes

The last three breakout sessions were dedicated to addressing the drafted approaches and subsequently developing the indicator categories. Workshop participants were also asked to investigate the responsiveness of proposed indicators to the NCA requirements for indicators and metrics. The workshop participants broke into four randomly selected breakout groups to address the drafted NCA indicator approaches (statistical, thematic, physical/system; see Box 4). Questions posed to the groups are listed in the agenda found in Appendix A. The breakout groups unanimously agreed that no one approach drafted by the NCA could be deemed comprehensive with respect to the requirements of the NCA vision. Most participants felt that the division between statistical, physical, and thematic approaches was rather arbitrary, and that a combination of all approaches may work better than any one approach alone. However, there was concern that a mix-and-match type approach could lead to more indicators than sought according to the NCA vision. In an effort to more fully expand upon the different concepts, the workshop participants were encouraged to continue development of their preferred approach (of those drafted) in subsequent breakouts. Two groups investigated the thematic approach, one the physical approach, and the last the statistical approach.

Draft NCA Approaches

The second breakout session was designed to provide feedback on the drafted NCA approaches. A number of helpful comments were provided by the workshop participants and are summarized in this section (Box 5). The statistical approach focused on the use of common statistical measures to organize possible indicators of climate change and impacts. One group considered the statistical approach best due to its flexibility, scientific foundation, and ability to represent changing conditions. All groups felt that the statistical approach was in some way embedded implicitly in the production of any indicator suite. That is, the statistical approach may be considered more as a tool to calculate indicators whereas the other approaches (physical and thematic) are more appropriate for providing context. It should be noted that the group focusing on the statistical approach questioned if the statistical approach could be better represented by organizing the various

measures (mean, variance, extremes, *etc.*) around more broad water- or temperature-based themes. This type of organization was discussed in other groups as well, and the possibility of a hierarchical approach was suggested that recognizes the basic data processing capabilities encompassed by the statistical approach, but also included other higher-level considerations.

Another common observation related to the statistical approach was the need to refine the terminology of some indicator categories. For example, the mean state is better described as an "expectation of normal" and variance can be couched as "range of possibilities." Several groups noted that the statistical measures important to the audience (*i.e.*, public) were often derivatives of a small set of state variables such as temperature, precipitation, streamflow, *etc.* The various metrics often require partitioning of these basic data using characteristics of amplitude, timing, and duration. Workshop participants noted that units of indicators should be made comprehensible to users. The numbers of hazardous events, the first day of some occurrence (*e.g.*, snow melt), or the length of an event (*e.g.*, in days) such as the growing season likely resonate well with personal experiences in terms of amplitude, timing, and duration.

The physical systems-based approach would partition climate change indicators based on large-scale features such as the atmosphere, the ocean, or the cryosphere, among others. Other indicator-based efforts such as from the EPA have focused on a similar physical systems-based organization. Most breakout groups noted that this organization could also be considered "thematic." They felt the distinction posed in the white paper between the physical and thematic indicator approaches was arbitrary. The individual categories within the physical system, though, can still be considered distinct from the statistical and thematic groups. That is, a system-oriented indicator category such as "Atmosphere" is distinct from a thematic indicator category such as "Water Cycle." The breakout groups differed on their opinion about the utility of the physical systems-based approach. The group focusing their attention on the physical system indicator categories felt the systems-based approach was appropriate because the categories would resonate well with an informed public. Contrarily, another group eliminated the physical approach in belief it did not relate directly to personal experience and therefore to the end user.

Box 5: Perspectives on the Drafted NCA Approaches

There was no single indicator approach (see Box 4) that resonated with all attendees. Some of the most frequent comments on the usefulness of the draft NCA approaches are listed below.

1. The statistical approach was implicitly embedded in any indicator suite.

2. The terminology of the indicators would need careful consideration for public communication.

3. The distinction between physical and thematic approaches was arbitrary.

4. The thematic approach seemed most appropriate for addressing policy and widelycommunicating with NCA audiences

A general organization of indicator categories using thematic types was selected as the most appropriate by two of the four breakout groups. The breakout group participants felt the use of themes would provide a higher-level approach capable of addressing policy questions and appealing to the broadest audience. They also thought the use of themes would be the most scalable between national and regional/sectoral levels. Interestingly, both groups focusing on the thematic approach attempted to derive indicators based on impacts to valued systems and resources. In essence, they tried to pose the question of what decision makers care about *versus* what climate scientists think are high priorities before identifying appropriate indicators. Those groups focusing on the thematic approach noted the need to go beyond the representation of basic variables and also stressed the need to include indicators based on metrics that are easily understood such as changes in extreme events and timing.

Proposed Indicator Categories
The four breakout groups each produced a set of potential indicator categories[4] that could be used by the NCA. Each of the groups followed a slightly different approach to identify indicators that would satisfy the requirements of the NCA vision. This

[4]Given the limited duration of the workshop, all potentially important indicators could not be discussed. Their exclusion from the set of proposed indicator categories does not indicate a lack of importance.

procedure has resulted in a number of possible indicators for monitoring the changing climate and its impacts. A list of the most fully developed indicators from each group has been organized in a manner so as to highlight commonality between the proposed indicators from the breakout groups (Table 2).

Several underlying themes were repeated across the breakout groups. The exact terminology differed between many of the groups but the basic concepts and data metrics were quite similar for many indicator categories. Four categories that were the most prevalent included extremes/natural hazards, a measure involving fresh water availability and quality, changes or large-scale shifts in patterns (*e.g.*, migration), and timing-related issues. It is to be expected that a number of metrics would be reused simply because there are a limited number of high-quality, spatially well-sampled observations. Their use in different indicator categories provides examples of how they may be repackaged to drive indicators providing important, distinct information. For example, "Food Production" contains timing metrics such as growing degree days. Yet it is combined with metrics such as soil moisture and precipitation to provide estimates on agricultural production. This summary attempts to capture the broadest results from the individual breakout sessions. Appendix B provides the details on individual discussion topics from each breakout group.

Thematic (a)	Thematic (b)	Physical	Statistical
Multiple References			
Human Health/ Hazards	**Extremes/Hazards**	Extremes/Hazards	Extremes/Hazards
Ecosystems	**Biophysical**	Land/Dryness	Migration
Water Quality/Quantity	**Hydro/Freshwater**	Lakes and Rivers	
	Timing	Warming Indicator	Timing
	Coastal	Oceans/Coasts	
	Snow/Ice	Snow/Ice	
Single Reference			
Food Production		**Greenhouse Gases**	**Mean States**
Energy Supply/Demand			**Variance**

Table 2. Indicator categories proposed by each of the four breakout groups have been organized such that similar indicators are found across each row. Categories referenced by multiple groups are listed under Multiple References while categories only mentioned by an individual group are listed under Single Reference. A missing entry implies that the listed group did not propose an analogous indicator. Note that similar indicator categories were proposed using differing terminology. In total, there are 11 indicators that stand distinct from one another. Bold face is used to indicate the term associated with these indicator categories.

Workshop Summary and Next Steps

Workshop participants provided useful insight into the development of a set of physical climate indicators to be used in the National Climate Assessment. Comments were made regarding the overall NCA framework, the development of useful indicator categories, and how to communicate those indicators to the target audience. Several important issues emerged through examination of the breakout discussions.

i. *There is a need to leverage existing work on indicators.* Several of the breakout groups noted a need to inventory the various indicator-based assessment efforts. As pointed out by one participant, the development of these indicators will likely rely heavily upon federal agencies; therefore it should be determined what is already at their disposal. Many felt that by addressing the question of what has been done, the NCA could potentially focus more concisely on the needs remaining to be addressed. Understanding the relationship to other indicator efforts will help the NCA develop indicators which are compatible with existing resources, an important constraint in a fiscally challenging environment.

ii. *Efforts must be made to relate climate change and impacts in a manner that people can readily perceive.* One participant noted that indicators act in two ways: to show that climate is changing and to show the impact of that change on human life. Individuals must understand how that change impacts them. It was noted that to communicate effectively with a broad audience, the NCA would need to go beyond the basic variables. Though indicator categories may be based on basic metrics of temperature and precipitation, it could be more useful to quantify derivatives from these metrics such as growing season length or number of days greater or less than some temperature threshold. Essentially, indicators must be something people can directly experience in their daily lives.

iii. *Indicators must be developed using high-quality data sources.* Indicators must satisfy the constraints of credibility and transparency. The NCA is already under much scrutiny. Results from these indicators must stand up to a thorough inspection of data sources and therefore data quality is a priority concern. Workshop participants agreed that data sources should be well-documented and peer-reviewed to the extent feasible. Using only the highest quality data sources may restrict what metrics are available for the construction of the indicators.

iv. *The NCA needs to work on clarification of its vision.* There seemed to be confusion among the workshop participants about the NCA vision (*i.e.*, purpose, audience, and scope). Several comments were made regarding the ability of the NCA to be, as one participant said, "everything to everyone." The NCA must strike a balance between providing overly detailed information at regional/sectoral levels at the expense of useful background information on a national scale. The NCA also needs to better identify the components of climate change it desires to assess. There was confusion as to how much the Assessment should slant towards climate change *versus* towards vulnerability and adaptation.

v. *Communicating the indicators will require careful consideration.* A number of presentations and breakout discussions expounded on the importance of crafting how to communicate indicators. This includes both the selection of units and the graphical display (*e.g.*, the use of maps is better than an x-y plot) of indicators. Another specific comment addressed the need for the NCA to engage communications specialists and graphic artists early in the process. Feedback needs to be actively sought from non-scientists on the intelligibility of the indicators to the general public.

A major expectation of this workshop was the production of a set of indicator categories for consideration by the NCADAC. Several indicator categories appeared as common threads across the breakout group discussions. Extremes/hazards and biophysical attributes were identified by all four groups as broad categories for physical indicators, hydrologic/freshwater and timing were identified

Indicator Category	Number of Groups	What do people care about?	
Extremes/Hazards	4	Flooding	Wildfire
		Drought/Heat Waves	Air Quality
		Tornadoes/Severe Weather	Temperature/Precipitation
Biophysical	4	Movement of Species	Desertification
		Temperature Zones	Fall Color
		Plant Hardiness Zone	Ice Band/Snow Cover
Hydrologic/Freshwater	3	Water Supply	Irrigation
		Water Quality	Soil Moisture
		Lakes and Rivers	
Timing	3	Growing Season Length	Length of Hurricane Season
		Heating/Cooling Degree Days	Phenology (first bloom, leaf-out)
		Ice-in/Ice-out	First and last (heat wave, freeze)
		Snow Melt	Peak run-off
Coastal	2	Sea Level	Ocean Acidification
		Sea Surface Temperature	Saltwater Intrusion
		Erosion	
Snow/Ice	2	Snow Cover Extent and Volume	Permafrost Extent and Volume
		Sea Ice Extent and Volume	Glacier Extent and Volume

Table 3. The most common potential indicator categories are listed, ordered by the number of groups that referenced a particular type of indicator category. For each category, several items that represent metrics/purposes that should be considered are presented. Some of these items could serve as an individual metric (e.g., length of hurricane season) while others may themselves entail a combination of metrics (e.g., severe weather).

by three groups, and coastal and snow were identified by two groups (Table 3). The overlap of broad categories among the groups reflects a degree of congruence about the central question posed to the groups – what do people really care or worry about in the physical climate system that could have a major impact on the earth system and human society in the future? However, the workshop was not intended to reach consensus, nor to resolve how indicators for these categories should be designed. Sustained research efforts will be required to craft the overarching framework for indicators prior to requesting data in any given indicator category. Research teams working in conjunction with the NCADAC will need to identify relevant topics and datasets, to experiment with different approaches to combining the metrics, and to carefully consider the presentation of the entire decision process for peer and public review.

Although there was overlap in the indicator categories among the groups, there remained a number of outstanding issues about how to approach the identification of physical climate indicators.

1. Should the portfolio of physical climate indicators include anthropogenic forcing (e.g., atmospheric greenhouse gas concentrations, land cover, or aerosols) or focus on the response of the physical climate system to forcing (i.e., the impacts of the forcing)?

2. What is the appropriate balance between easily measured and consistently quantified physical changes in the climate system (e.g., global sea-level change, mean annual minimum or maximum temperature, seasonal precipitation) versus regional or sectoral impacts of physical changes in the climate system (e.g., temperature extremes and shifts in the timing and magnitude of seasonal changes)?

3. Is the role of physical climate indicators to provide information on the past, present, and future behavior of the climate system or should indicators be designed to directly address critical societal and environmental

vulnerabilities and adaptation responses to changes in the physical climate?

4. Finally, while noting that the workshop charge was to identify a set of broad categories for physical indicators to assess the changing climate and potential impacts on national interests, the sense among a significant number of the participants was that physical indicator development aimed at supporting adaptation was more compelling. However, a problem-focused approach resulting in the co-production of a suite of a truly integrated ecological, physical, and societal climate indicators that could meet the collective needs of the regional, sectoral, and national climate assessments was beyond the scope of the physical climate indicators workshop.

As noted in the introduction, the physical climate indicators workshop was the second in a series of three workshops intended to acquire community input on how the NCA could develop a suite of climate change indicators as part of its ongoing assessment process. The ecological indicators workshop took place in late November 2010, and the societal indicators workshop occurred in late April 2011. Once reports from all three workshops are complete, an obvious next step will be to blend the report findings and knowledge gained into an overall national indicators structure that includes an integrated suite of ecological, physical, and societal climate indicators that meets the needs of the NCA.

The expectation is that the NCADAC will create a working group comprised of FAC members and other representatives who would first work to clarify and finalize the overarching NCA indicator vision. This clarifying effort will be necessary to resolve many of the quandaries and divergent views raised by the participants of all the workshops. **First order issues include how to build a relatively simple indicator list while still encompassing important climate change impacts, vulnerability, and adaptation; a high degree of coordination across ecological, physical, and societal indicators; and capturing and conveying sufficient information richness in the indicator suite for all regions and sectors of the Nation.**

Appendix A: Agenda

Tuesday, March 29, 2011

8:30   Introduction and meeting logistics - Fred Lipschultz (USGCRP)

8:35   Host welcome - Jack Kaye (NASA)

8:50   The National Climate Assessment and Indicator Vision - Kathy Jacobs (OSTP)

9:20   White Paper summary and products from the meeting - Ken Kunkel (Cooperative Institute for Climate & Satellites, NOAA)

9:40   Q&A - NCA Staff

10:00   **Panel Discussion:** *Applications and Uses of NCA Physical Climate Indicators*

Rick Murnane (Bermuda Institute of Ocean Sciences)
"Everybody talks about climate change, but do (re)insurers do anything about it?"

Steve Gray (Wyoming State Climate Office)
"Climate Indicators for Use in Natural Resource Management and Planning"

David Robinson (Rutgers University)
"Stakeholder perspectives on indicators"

Facilitated discussion with panel of speakers – Kathy Jacobs

11:15   **Panel Discussion:** *Approaches to Developing Physical Climate Indicators*

Tom Karl (USGCRP, NCDC)
"Developing Climate Indicators --- Too few, too many, or just right?"

Steve Running (University of Montana)
"IGBP Climate Change Index"

Jim Butler (NOAA Global Monitoring Division)
"NOAA Annual Greenhouse Gas Index"

Mike Kolian (EPA)
"EPA's National Climate Indicators"

Facilitated discussion with panel of speakers – Robin Webb (NOAA)

1:15   **Breakout:** *The NCA Indicator Vision (See White paper Sections 1 and 2): Comments for improvement*

Breakout group charge - Kathy Jacobs & Melissa Kenney
- Is the proposed purpose for the NCA indicators clear, complete and compelling, and what changes might improve these?
- Scope: How should the physical climate indicators relate to regions or sectors for the NCA? How should the physical climate indicators relate to ecological and societal indicators of climate change also being developed for the NCA?

- Audience: Are the potential user groups of the NCA indicators appropriate, or too narrow or broad? Do the user group's needs and indicator purpose align?

1:30   Move to breakouts.

3:00   **Breakout:** *The draft NCA Category Approaches: Are these appropriate and are there others to consider?*
        Breakout group charge - Fred Lipschultz
        - Which indicator category approaches in the white paper (Section 3) provide the most meaningful and informative approach; are there other approaches that should be considered?
        - Does the candidate category of indicators satisfy the NCA criteria of audience, scope and purpose?
        - Can the categories be transferred to ecological and societal indicators?

3:15   Move to breakouts. Groups will each address the following questions:

4:30   Breakouts return and Facilitators briefly report back on approaches.

5:00   Adjourn

Wednesday, March 30, 2011

8:30   Welcome – Jim Buizer (University of Arizona)

8:35   Synthesis of potential categories for indicators from Day 1– Fred Lipschultz

9:00   **Breakout:** *What are the adequate & appropriate metrics/variables for each indicator?* (White paper **Section 4**)
        Breakout group charge - Fred Lipschultz:
        - Brainstorm potential metrics/variables for the indicator
        - Apply the NCA framework for indicator properties to each metric/variable
        - Discuss which metrics/variables best inform the indicator
        - How might the variables be combined in such an indicator; *e.g.*, normalizing trend analysis

9:10   Move to breakouts. Groups based on results from morning discussion.

11:15   **Speaker: Peter Colohan (Senior Policy Analyst, OSTP)**
        "A national observing strategy for the USGCRP: data streams for the NCA indicator suite"

        Q&A/Discussion facilitated by Allison Leidner (NASA)

1:15   Report from morning breakout sessions - progress, questions, issues, successes

1:30   **Continue Breakout:** *What are the adequate & appropriate metrics/variables for each indicator?* (White paper **Section 4**)
        Breakout group charge from morning session (refine as needed) - Fred Lipschultz
        - Which metrics/variables best inform the indicator?
        - How might the variables be combined in such an indicator analysis?
        - What data and observing assets are available? Are there significant data gaps that can be identified?

1:45   Move to breakouts

3:00   Speaker: Roger Pulwarty (NOAA)
          "Connecting the NCA Indicator suite to the IPCC and other climate risk assessments"

3:30   Facilitated Discussion:
          Moving Forward: Strategy, Priorities, and Identifying Resources - Jim Buizer

4:45   Final thoughts - Jim Buizer

5:00   Adjourn

Appendix B: Breakout Group Summaries

The development of the indicator categories was structured around several breakout sessions and included the division of workshop participants into four groups. On Day 1 of the workshop, the breakout groups were randomly selected. Three preferred approaches emerged from discussion by these groups: Statistical (1 group), Physical (1 group), and Thematic (2 groups). To stimulate development of these ideas, it was requested that these approaches serve as the focus of breakout groups for the Day 2 discussion. Workshop participants were allowed to self-select the group in which they participated on the second day. The following summaries provide additional details of the discussion from these groups.

Statistical

This group noted five possible indicator categories (mean states, timing, migration, extreme events, and variance) but only made significant progress on the development of an extreme events category. Indicator metrics for the Extreme Events category included: heavy precipitation, air stagnation, severe weather, heat waves, flood incidence, drought incidence, tropical cyclones, and killing frosts.

Regarding the need to aggregate metrics, the group questioned whether a composite indicator of extremes is more meaningful to a user or whether the NCA should simply track the number and type of extreme events. The group also proposed a small set of metrics for the Migration category including: shift in temperature patterns, shift in precipitation patterns, the shift of plant hardiness zones, snow cover, desertification, glaciers, coastal location, and ice accumulation bands. However, they did not formally evaluate these metrics against the NCA requirements. Discussion of the other categories remained limited. This group also discussed the need to provide an inventory of the available indicators. They felt that an inventory might make it easier to determine the indicators on which to focus by examining what is currently lacking. This group spent much time debating the semantics of the properties of metrics for each indicator. As an example, they discussed whether the "ambiguity" of a metric referred to the ambiguity of the metric itself or to the ambiguity of changes in the metric over time.

Physical

The group that explored the physical systems approach discussed seven potential indicator categories. For each of these categories, this group examined individual indicators that could be placed within each category. Some requirements placed on indicators by this group included i) the availability of high-quality data; ii) the ability to be geographically represented; and iii) the ability to resonate with the public. The seven categories and some proposed indicators are listed here:

1. Extremes Indicator: U.S. land-falling hurricanes, precipitation extremes, heat wave index, cold wave index, nor'easters, storm surge, severe thunderstorms, hail, tornadoes, and lightning
2. Warming Indicator: duration of growing season, trends in water and soil temperatures, heating and cooling degree days, lower and upper tropospheric air temperatures
3. Greenhouse Gases: radiative forcing component
4. Oceans and Coasts: sea level, sea surface temperature
5. Dryness: Palmer drought index, percent of dry/wet land, soil moisture and soil temperature, thermal response number
6. Lakes and Rivers: stream flow, lake levels, water temperature, lake turnover
7. Snow and Ice: extent of snow cover, arctic sea ice, and permafrost, timings of ice in/out, melt season, snow cover, and volume of snow pack

Thematic (a)

Of the two groups focused on a thematic approach, the group discussed here took a "human-needs" based approach to the development of the indicator categories. They felt any indicator category should be able to stand up to the "so what" question that could be posed by the public. In addition to the general importance of an indicator, this group imposed additional constraints on indicators. They need to be scalable, able to be integrated nationally, and based on high-quality observations. This group developed a set of five indicator categories. For each category, there were additional components (e.g., indicators) that could be developed from a set of metrics. The following list provides the hierarchy or categories, indicators, and metrics developed by this group:

1) Fresh Water
   a. Supply
      i. Lake levels, reservoir levels, snow cover, snow pack, groundwater, precipitation, stream flow,
         imports/exports
   b. Quality
      i. Salinity, nitrogen, precipitation, water temperature, dissolved oxygen
2) Food Production
   a. Agricultural Crops
      i. Temperature, water supply, precipitation (type, amount, intensity), growing degree days, soil
         moisture
   b. Fish/Aquaculture
      i. Salinity, dissolved oxygen, temperature
   c. Livestock
      i. Temperature, precipitation
3) Human Health
   a. Non-Hazard
      i. Temperature, air quality, water quality, water quantity, precipitation, humidity
   b. Hazard
      i. Storm surge, precipitation, sea level, wind temperature, severe weather, tornadoes, hurricane
         intensity and frequency, lightning, wildfire
4) Ecosystems
   a. Terrestrial
      i. Temperature minimum/maximum, Precipitation type/quantity, solar radiation, growing degree
         days, albedo, extreme weather, fire, evapotranspiration, soil moisture/drought, permafrost
   b. Freshwater
      i. Water temperature, air temperature, water quality, water supply, streamflow
   c. Marine
      i. Water temperature, water quality, sea level, salinity, acidity, currents, solar radiation
5) Energy
   a. Supply
      i. Solar insolation, wind speed and persistence, temperature, precipitation, waves
   b. Demand
      i. Temperature, humidity, precipitation, extreme weather

Thematic (b)

In a similar approach to the other thematic group, this group formulated the indicator categories by answering the question "Why do people care?" This group felt that indicators should relate to personal experiences. This group recognized that many indicators would necessarily be interdependent when attempting to develop broad, high-level indicators. This group leaned towards a 'vital signs' framework for the indicator categories rather than a 'warning lights' framework. This group also stressed the need to have high-quality data for any indicators. Each of the six categories proposed by this group contained indicators and metrics driven by an intent to inform a specific concern under each category. The outline of these indicators are listed here:

1) Hydro
   a. Water Supply
      i. Depth of aquifers, reservoir capacity, naturalized stream flow, peak run-off
   b. Irrigation
      i. Daily precipitation minus daily potential evapotranspiration
   c. Soil Moisture
      i. Saturation, percent water content
   d. Lake and River Levels
2) Biophysical (ecological, growing season)
   a. Fall Color
   b. Appearance/Disappearance of a Species
3) Timing
   a. Length of Hurricane Season
   b. Peak run-off
      i. Duration and timing of peak and low-flows
   c. First and Last Freeze
   d. Ice-in/Ice-out
   e. Monsoon Dates
      i. Duration, number of hours above dew point 55°F
   f. First Snow Cover
   g. Phenology (first bloom, leaf-out)
      i. Days above or below a certain temp threshold
   h. First and Last Heat Wave
4) Coastal
   a. Loss and Damage of Natural and Built Environment
      i. Inundation, erosion
   b. Coastal/Ocean Productivity
      i. Ocean acidification
   c. Coral Bleaching
      i. Maximum temperature exceeded

       d. Saltwater Intrusion
          i. Brackish conditions, salinity
5) Extremes/Hazards
       a. Heat Stress and Mortality
          i. Heat Index
       b. Flooding
          i. Frequency and rate of heavy
          precipitation, rapid snow melt, 1-in-20
          year events, *etc.*
       c. Hurricanes/Tropical Cyclones
          i. Frequency, magnitude, timing, and
          extent
       d. Winter Storms
          i. Frequency, magnitude, timing, and
          extent
       e. Drought
          i. Daily precipitation minus daily potential
          evapotranspiration
6) Snow/Ice Cryosphere
       a. Sea Ice
          i. Extent, duration, thickness, fast *vs.* pack
          ice, first-year *vs.* multi-year ice
       b. Frozen Ground
          i. Depth, duration
       c. Lake and River Ice
          i. Ice-out, ice thickness
       d. Snow on-the-ground
          i. Extent, duration, daily/weekly/monthly,
          liquid water equivalent and depth
       e. Glaciers
          i. Size, extent, and total mass

Appendix C: Workshop Planning Committee

Muthuvel Chelliah
NOAA
Muthuvel.Chelliah@noaa.gov

Emily Cloyd
U.S. Global Change Research Program
ecloyd@usgcrp.gov

Kathy Jacobs
White House Office of Science and Technology Policy
Katharine_L._Jacobs@ostp.eop.gov

Randy Johnson
U.S. Department of Agriculture
randyjohnson@fs.fed.us

Ken Kunkel
NOAA
ken.kunkel@noaa.gov

Allison Leidner
NASA
allison.k.leidner@nasa.gov

Fred Lipschultz
U.S. Global Change Research Program
flipschultz@usgcrp.gov

Jeff Luvall
NASA
jluvall@nasa.gov

Christa Peters
NASA
christa.peters@nasa.gov

Brent Roberts
NASA
jason.b.roberts@nasa.gov

Jonathan Smith
U.S. Geological Survey
jhsmith@usgs.gov

Jim Smoot
NASA
james.l.smoot@nasa.gov

Robert Webb
NOAA
Robert.S.Webb@noaa.gov

Ken Wolfenbarger
NASA Jet Propulsion Laboratory
james.k.wolfenbarger@nasa.gov

Appendix D: Workshop Participant List

Deke Arndt
NOAA National Climatic Data Center
Derek.Arndt@noaa.gov

Jeff Arnold
U.S. Army Corps of Engineers
Jeffrey.R.Arnold@usace.army.mil

John Baker
U.S. Department of Agriculture
john.baker@ars.usda.gov

Sandra Baptista
Center for International Earth Science Information
Network
Columbia University
sandra.baptista@ciesin.columbia.edu

Levi Brekke
U.S. Bureau of Reclamation
LBrekke@usbr.gov

James Buizer
Arizona State University
jim.buizer@asu.edu

Jim Butler
NOAA Earth System Research Laboratory
James.H.Butler@noaa.gov

Ralph Cantral
U.S. Global Change Research Program
rcantral@usgcrp.gov

Muthuvel Chelliah
NOAA Climate Prediction Center
Muthuvel.Chelliah@noaa.gov

Chris Clark
Environmental Protection Agency
clark.chrism@epa.gov

Emily Cloyd
U.S. Global Change Research Program
ecloyd@usgcrp.gov

Peter Colohan
White House Office of Science and Technology
Policy
peter_e._colohan@ostp.eop.gov

Joey Comiso
NASA Goddard Space Flight Center
josefino.c.comiso@nasa.gov

Tom Dinardo
U.S. Geological Survey
tpdinardo@usgs.gov

John Dwyer
U.S. Geological Survey
dwyer@usgs.gov

Dennis Dye
U.S. Geological Survey
ddye@usgs.gov

David Easterling
NOAA National Climatic Data Center
David.Easterling@noaa.gov

Carolyn Enquist
University of Arizona
cenquist@email.arizona.edu

Lawrence Friedl
NASA Headquarters
Lawrence.A.Friedl@nasa.gov

Chelsea Friedman
NOAA Climate Program Office
chelsea.friedman@noaa.gov

Bryce Golden-Chen
U.S. Global Change Research Program
bgoldenchen@usgcrp.gov

Patrick Gonzalez
U.S. National Park Service
patrick_gonzalez@nps.gov

Steve Gray
University of Wyoming
sgray8@uwyo.edu

Dorothy Hall
NASA Goddard Space Flight Center
dorothy.k.hall@nasa.gov

David Halpern
NASA Headquarters
david.halpern@nasa.gov

Peter Hildebrand
NASA Goddard Space Flight Center
peter.h.hildebrand@nasa.gov

Christina Hudson
Science Applications International Corporation
christina.c.hudson@saic.com

Kathy Jacobs
White House Office of Science and Technology
Policy
Katharine_L._Jacobs@ostp.eop.gov

Randy Johnson
U.S. Forest Service
randyjohnson@fs.fed.us

Alexey Kaplan
Columbia University
alexeyk@ldeo.columbia.edu

Tom Karl
NOAA National Climatic Data Center
Thomas.R.Karl@noaa.gov

Jack Kaye
NASA Headquarters
Jack.A.Kaye@nasa.gov

Melissa Kenney
NOAA Climate Program Office
melissa.kenney@noaa.gov

Michael Kolian
Environmental Protection Agency
kolian.michael@epa.gov

Ken Kunkel
NOAA National Climatic Data Center
ken.kunkel@noaa.gov

Ron Kwok
NASA Jet Propulsion Laboratory
ronald.kwok@jpl.nasa.gov

Allison Leidner
NASA Headquarters
allison.k.leidner@nasa.gov

Maxine Levine
U.S. Department of Agriculture
Maxine.Levin@wdc.usda.gov

Fred Lipschultz
U.S. Global Change Research Program
flipschultz@usgcrp.gov

Jeff Luvall
NASA Marshall Space Flight Center
jluvall@nasa.gov

Molly Macauley
Resources for the Future
macauley@rff.org

Rick Murnane
Bermuda Institute of Ocean Sciences
rick.murnane@bios.edu

Sheila O'Brien
U.S. Global Change Research Program
sobrien@usgcrp.gov

Toral Patel-Weynand
U.S. Forest Service
tpatelweynand@fs.fed.us

Christa Peters-Lidard
NASA Goddard Space Flight Center
christa.d.peters-lidard@nasa.gov

Rich Pouyat
U.S. Forest Service
rpouyat@fs.fed.us

Roger Pulwarty
NOAA Climate Program Office
Roger.Pulwarty@noaa.gov

Brent Roberts
NASA Marshall Space Flight Center
jason.b.roberts@nasa.gov

David Robinson
Rutgers University
drobins@rci.rutgers.edu

Matthew Rodell
NASA Goddard Space Flight Center
Matthew.Rodell@nasa.gov

Steve Running
University of Montana
swr@ntsg.umt.edu

Arthur Rypinski
U.S. Department of Transportation
Arthur.Rypinski@hq.doe.gov

Steve Sebestyen
U.S. Forest Service
ssebestyen@fs.fed.us

Jim Smoot
NASA Marshall Space Flight Center
james.l.smoot@nasa.gov

Anne Waple
NOAA National Climatic Data Center
Anne.Waple@noaa.gov

Chris Weaver
Environmental Protection Agency
weaver.chris@epa.gov

Robin Webb
NOAA Earth System Research Laboratory
Robert.S.Webb@noaa.gov

Ken Wolfenbarger
NASA Jet Propulsion Laboratory
james.k.wolfenbarger@nasa.gov

www.ingramcontent.com/pod-product-compliance
Lightning Source LLC
Chambersburg PA
CBHW081416170526
45166CB00010B/3367